# A TEACHER'S GUIDE TO
# MATHS AND THE HISTORIC ENVIRONMENT

Tim Copeland

English Heritage

# CONTENTS

**Maths project work at Kenilworth Castle, Warwickshire.**

# ABOUT THIS BOOK

This book has been written with the aim of bringing maths to the forefront of teachers' and children's thinking about the past and the physical evidence that survives from past times. Although it might appear at first sight to be about those preserved monuments that we take children to visit as part of a project on history, it is really about all the structures from the past that surround children in their everyday lives.

The book is as much about maths as it is about historical sites. Maths that is taught without reference to it's power to explain real problems is a stilted maths, like learning to speak without ever having a conversation. To be successful in our teaching of maths we need to engender in children a feeling for the unique problem-solving aspects of mathemematical processes, and their power to aid explanation. Any place could be the focus for using mathematical skills for their own sake. The maths in this book have been chosen because they add knowledge and enjoyment of particular types of site.

The Cockcroft Report into the teaching of maths pointed to the unique role of the subject as a powerful means of communication. The HMI 'History 5-16' document suggests that one of the reasons for studying the past is 'to develop insight, clearly based on historical evidence, in order to offer explanations of past events and to develop also an informed appreciation of the perspectives and motives of people in the past.'

What more suitable place to use maths than in communicating with those who lived in the past by solving the problems left us by now silent peoples, who recorded little, but left an abundance of clues for us to use?

With the coming of the National Curriculum and the publication of the Orders for Mathematics, there is now a set minimum syllabus for maths, and this book has had that development in mind. However, it is not possible to do justice to the maths of a site by just referring to the Attainment Targets in the Mathematics Orders, and where tasks suggested fall outside the statements of attainment, they will surely contribute to a wider use of maths, and to other subjects, English, Science, History, or Geography. Maths is not the sort of tool that can be used just for itself, it is powerful because it permeates many other areas of thought.

# WHAT IS IT ALL ABOUT?

Maths and the historic environment seem like a strange combination. Doesn't maths have to do with number and logic, and history, isn't that to do with people and places in the past? Isn't it a bit too complex a pairing for children to be expected to use? Maths and the historic environment are intimately connected in many ways, as this book will show.

## WHAT IS THE HISTORIC ENVIRONMENT?

In its widest sense the historic environment is all the elements from the past that surround us. So, the landscapes and townscapes we are familiar with, being the result of human activity in the past, are all part of the historic environment. Wherever we look we are confronted by the intended, and incidental, influence of human beings. Completely natural vistas are becoming increasingly rare in the Britain, although we tend to associate the woodlands, fields and high moorlands with the natural world, they are all the result of the modification of that natural world by human beings. The river valleys and low undulating and flat areas have been altered most, but even the high tops of the mountains in the upland areas have been influenced by grazing activity of sheep and cattle, and more recently the hiking boot, and are no longer in any 'natural' state.

For the purpose of this book the historic environment will mainly highlight structures built to satisfy human needs - home, farm, village, fields, buildings for defence, worship, manufacture. This definition is still wide considering the length of time that these islands have been inhabited, and the number of different types of structure that have been built. However, if wide, it allows room for variety and diversity, and enables those interested to have an exciting time working with the historic environment.

Maths is concerned with a way of thinking, a logical progression of concepts and skills for solving a variety of problems, and uses a large number of thinking processes besides the formal computation of numbers. Planning, making sense of space, decision taking, using pictorial representations to aid understanding, measuring, and searching for similarities and differences, are some of the ways that maths is used in everyday life, just as they were in the past. Because it is a powerful method of communicating information, maths has an important role in answering many of the questions posed about the historic environment.

## HOW DO CHILDREN ENCOUNTER THE HISTORIC ENVIRONMENT?

For the teacher wishing to use first hand experience in teaching, this is a fundamental question. The obvious answer is that they encounter it every day in every moment of their lives, and perhaps have a keener sense of the past than most adults, since what teachers remember as familiar because it happened in their lives, for children with much shorter lifetimes, nearly all

ENGLISH HERITAGE

that is around them is the product of a 'distant' past.

■ Intact standing structures: schools, churches, town halls, cathedrals, castles, all intact and used for a present day purpose, are the most frequently met remains of the past.

■ Incomplete standing structures: no longer used because they have been subject to the activities of human beings in being damaged, destroyed, and along with biological agencies, such as rot, and weathering, have been left ruined. These are the sites most frequently visited by school children in the course of their work. Often the destruction has been intense, and the evidence is now in the form of earthworks, banks and ditches, and multi- shaped mounds.

■ Buried structures: often monuments, have disappeared altogether. The

reasons are because they were made of perishable materials, wood for example, because their materials were looted and used elsewhere, or because their standing remains have been buried by other activities. All these structures leave some trace, and it is the job of the archaeologist to discover such structures and to make sense of what is left. Often excavated sites, now preserved, are visited by school children. Children will often, though, have the opportunity to view objects from excavated sites in museums. Of course many monuments fall into each category with elements upstanding and buried.

Children visit parts of the historic environment as part of their studies, and often these sites are those preserved, and maintained, by such bodies as English Heritage. These are sites of national importance, and are preserved for future generations, as well as those alive today. Many sites of national

importance will be preserved in the area of most schools, but are not open to the public. These are Scheduled Ancient Monuments, because they are on the 'Schedule' of protected monuments. However, there are many sites of local importance that are not Scheduled, but are equally as old, and there are many buildings which are important in terms of the children's personal heritage: home, school, the road or street, which also form part of the historic environment and can be usefully studied.

The list of English Heritage monuments (see Bibliography) gives some idea of the range and diversity of sites to be seen in our landscape and towns. Perhaps you can add a few suggestions from your own locality. It is useful to ask children for their own examples, as you will begin to get an insight into the what children consider to be part of the past.

**Intact: Osborne House, Isle of Wight — Queen Victoria's seaside home.**

**Incomplete: Wroxeter Roman City, Shropshire — some of the walls of the town's bath-house are now preserved above ground.**

**Buried: Aerial photograph of the deserted medieval village of Thwing in North Yorkshire. Most of the village's remains lie under grass which has produced parch marks. At the bottom left the house foundations and tracks show as disturbances in the ploughed field.**

**INTACT**

**BURIED**

**INCOMPLETE**

DOMINIC POWLESLAND/ENGLISH HERITAGE

# MATHS AND THE HISTORIC ENVIRONMENT

When we use the historic environment in the curriculum we are asking a series of questions, although they are often intuitive and unspoken. The main question is: 'What was this place like, and how has it changed?' That leads to a further series of questions: 'What happened here?', 'How old is it?', 'Who built it?', 'Why did they build it?', 'How did they build it?', 'How is it connected with other, similarly old places?', 'What happened here?', 'How do we know?'. Each of these questions has a common impulse behind it. We want to make sense of the physical evidence that we see, with the eventual aim of knowing what it was like to be there when the structure was in use.

Maths can help us answer these questions in many ways:

■ it can help us get to grips with the process of time passing, and put events and the structures associated with them, into perspective;

■ it can help us understand how maths was used in building that environment, and how the maths developed, because the maths often developed according to need;

■ it can help us record and represent the remains of structures from the past;

■ it can help us interpret and explain how those structures were used.

ENGLISH HERITAGE

Teaching anything successfully is about getting children to do the thinking, and children need to have practical activities to help them make sense of the mathematical concepts we wish them to know and use. There needs to be concept and skill getting and also concept and skill using. The relationship is not a straight one because 'getting' is enhanced by 'using', and 'using' often leads to a further 'getting' of skills and concepts. We need to provide situations where children will want to use the maths because they are interested enough to want to solve a meaningful problem. The historic environment provides many problems that are of intense interest to children because it attracts their sense of wonder.

**Facilitates**

**Skill getting** **On site** **Skill using**

**Motivates**

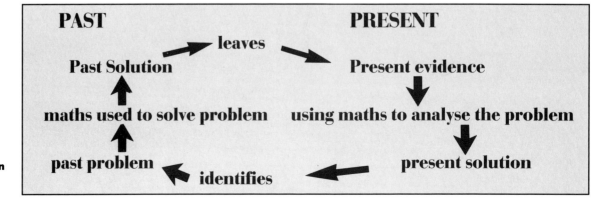

**PAST**

**PRESENT**

Past Solution → leaves → Present evidence

maths used to solve problem ← using maths to analyse the problem

past problem ← identifies ← present solution

**Problem-solving can be used to find how people in the past lived using this cycle.**

## HOW DOES THE MATHS FIT IN?

In trying to set up activities to be used when we visit an historic site with children we usually 'brainstorm' the possibilities, and see how they connect to the study we are undertaking at any particular time.

### Pose the problem:

Which site? How is it related to the topic being undertaken by the children? What types of structure are there? What skills, concepts and attitudes do we want to develop?

Once we have the broad outlines of our intentions, and the site is chosen, we are able to categorise our ideas in a topic web.

### Outline the questions

Efficient problem-solving is concerned not only with having an interest in something and the motivation to see the solving through, but the ability to pose questions. It is difficult to find focused answers without key questions.

ENGLISH HERITAGE

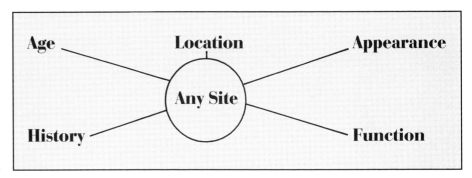

**Refine into areas for investigation.**

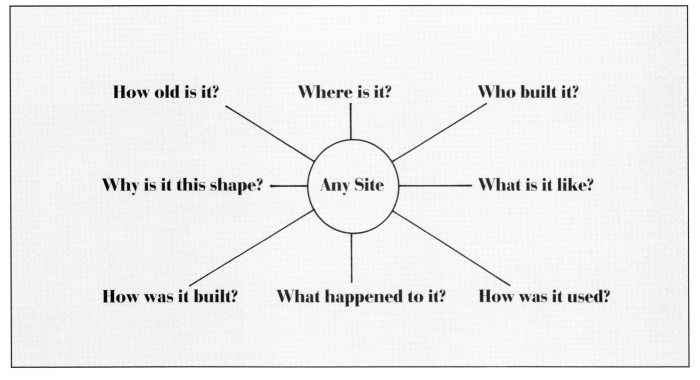

## People and places

With the questions in mind it is possible to set up a range of key activities that will provide for their solutions using all the facets of the site being visited.

When a question is asked this immediately is followed by others: What maths do we need to use? How are we going to use it? Where are we going to use it? Do we need to learn a new skill before we visit the site? What equipment do we need? This stage is where the appropriate mathematical skill has to be selected, and then used, and the results recorded as a picture, sentence, plan, table, or matrix, for example.

## Look for a solution

The maths is a means to finding an answer to the question set, but an important part of the process will be discussing the results to see if an appropriate solution has been found. By 'appropriate' we mean sensible or possible. 'Correct' answers can rarely be found on historic sites since we cannot get into the minds of the original users, so the children's ideas must be valued if they correspond to the available evidence.

## SO WHAT NEXT?

This is an opportunity to look backwards and forwards. The teacher might assess the children's learning against the criteria of the National Curriculum, or their own role in the experience. It is also a chance to plan for the future using any particular interests that the children might have shown, or any skills or concepts that they need to develop.

The important part of the whole process is asking the right questions and using the appropriate maths to get the data needed. The remainder of this book will be concerned with asking questions and choosing the appropriate mathematical processes to use to find solutions and to represent data.

The key activities vary in difficulty as they have been designed for the age spread 5-16 years. Some of the more complex tasks can be adapted for use by younger children, whilst the simpler ones can be used as a starting point and extended for older children. In the early years the teacher will have a substantial input and will have to suggest the questions and the maths needed to find an answer. Children in the middle years should begin to suggest the maths themselves, whilst older pupils might be able to design their own questions and find solutions to problems within the broad areas of siting, measurement, age, and shape of the monuments.

## BACK IN THE CLASSROOM

Back in the classroom, or in the Education Centre on a site, it will be possible to record the data collected in a more permanent form, drafting and redrafting the results for others to share, and adding diagrams, charts or graphs.

It is always satisfying for teacher and child to see a display, or an assembly, by putting together the results of each of the questions asked by different individuals or groups.

## WHO LIVED HERE?

Whenever we visit an historic site we try to 'people' it, and like to see artists' illustrations how it might have looked. This helps us put it into context and gives it a kind of familiarity. We are able to identify, in a limited way, with those who once lived there. The guidebook, in the case of a medieval site, might be able to name some of the owners of the monument, however those individuals are certain to have been high up the social scale. What about those nameless people who inhabited prehistoric sites,

or were in the lower ranks of medieval society? Maths cannot identify those hidden individuals, but it can be used very simply to demonstrate that the people who built and lived on many of the historic sites we visit are part of some people's own histories. The discovery that some of our ancestors actually lived on or near a particular site can increase motivation and children's sense of wonder. Family trees start with ourselves and our mothers and fathers. If we build our family tree downwards in time, we see that each of

us must have an increasing number of direct ancestors the further back we travel. The diagram below indicates the progression: 'I' stands for the individual, 'f' for father, 'm' for mother, 'g' for grand- or great-, 'ggf' is therefore great-grandfather, and 'ggm' great-grandmother and so on.

## GENERATIONS

| Generation | Number of people | | Generation | Number of people | |
|---|---|---|---|---|---|
| 1 | 2 | $2^1$ | 11 | 2,048 | $2^{11}$ |
| 2 | 4 | $2^2$ | 12 | 4,096 | $2^{12}$ |
| 3 | 8 | $2^3$ | 13 | 8,192 | $2^{13}$ |
| 4 | 16 | $2^4$ | 14 | 16,384 | $2^{14}$ |
| 5 | 32 | $2^5$ | 15 | 32,768 | $2^{15}$ |
| 6 | 64 | $2^6$ | 16 | 65,536 | $2^{16}$ |
| 7 | 128 | $2^7$ | 17 | 131,072 | $2^{17}$ |
| 8 | 256 | $2^8$ | 18 | 262,144 | $2^{18}$ |
| 9 | 512 | $2^9$ | 19 | 524,288 | $2^{19}$ |
| 10 | 1024 | $2^{10}$ | 20 | 1,048,576 | $2^{20}$ |

Each preceding generation has twice as many individuals in it as the generation that came after it. So if we count through the generations, starting with that of our parents, we can construct a table of our family population each generation.

So for each one of us 20 generations ago we would have had 1048 576 direct ancestors. If we take 25 years as a generation then in the year AD 1490, four years after Henry VII came to the throne, we had 1,048,576 direct ancestors living. Ten generations before that, we would have had 1,073,741,824, direct ancestors living, that is in AD 1240, when Henry III was on the throne of England.

The population of the world is not thought to be anywhere near this number at that time, and we have to take into account that inbreeding, with fairly close relatives marrying and perhaps few newcomers joining the local historic population. This means that these numbers are disproportionately large and that reductions should be made for the total number of ancestors. However, the fact that each of us must have hundreds or thousands of direct ancestors alive and living in different parts of the country before AD 1790, means that there is a good likelihood that at least one of our ancestors was living near the sites that we visit with children. It's a small historical world!

## Why did they choose this place?

Each historic site was located for a variety of reasons. Some of them might have been political, but the broad choice of site usually was dependent on its function: monasteries tend to be in out of the way places, castles in good defensive positions, country houses have good views.

The actual choice of a particular site may have been for reasons that are still very clear, and can be related to the children's everyday lives. The need for water, fuel, defence, communication are attributes that we share with those who lived in the past.

## What would you need to live in this place?

Children can soon name the essential physical necessities of a site, the attributes, and can look for the well or stream, the possible source of wood, the high position of the place. Perhaps the best way of recording is by the use of an ordered list, quantifying the priorites the chidren feel are important, or by drawing a sketch plan which can vary in the amount of information it contains according to the age of the children. At its most sophisticated, distance and direction could be included.

## How can we record the advantages and disadvantages of the site?

By using an attributes matrix children can decide whether they think the siting was good or bad, and possibly how the setting has changed. The important aspect of this type of enquiry is to let the children decide the attributes for themselves, with information about the use of the site being provided by the teacher.

## How did they decide?

In the immediate locality of the site being studied there will be other possible places that could have been used. These nearby sites can be examined and evaluated, again using an attributes matrix, to try to deduce why the site that was eventually developed was chosen, for example.

**Rievaulx Abbey, North Yorkshire.**

It may be that the children think that one factor is more important than the others, eg its siting on a hill, and can 'weight' this factor by 2 or 10, or whichever value is thought appropriate. thus giving a preference matrix. Similarly, the process can be used for individual parts of the site to ascertain why the structures were put in a particular place. On some sites, particularly defensive sites, the form of the landscape influences the shape of the site, or the strength of its defences. By constructing a matrix with the possible attributes of the nearby landscape - steep slopes, river, stream, boggy land, cliffs, gentle slope, flat land, - the obstacles in front of a particular feature - ditch, tower, gateway - can be quantified and the reasons for that particular structure can be discovered.

| | Good vision | high up | water | road | fuel | level ground |
|---|---|---|---|---|---|---|
| Site A | | | | | | |
| Site B | | | | | | |
| Site C | | | | | | |

# WHAT WAS IT USED FOR?

We can often identify the type of site that we are visiting (for example a castle, an abbey, a country house or a circle of stones) but the individual room or part of a structure takes a lot more detective work. Many of the questions this book suggest give partial information about the use of the component parts of a site. Now the information gained will be put together to try to find the uses of particular rooms. All sites, including our own homes and schools, have areas which have greater importance than other parts of the building. We can call these 'high status' areas, because they show visitors our wealth, sophistication or the ethos of the structure. Usually, these areas have specific attributes that indicate their importance, for example they are better decorated, more comfortable, take up more space, are

Buckminster County Primary School, Leicestershire. Built in 1899 for the Earl of Dysart as 'The Buckminster Unsectarian School'.

JILL MCPHERSON

better heated and lit, than the 'low status' areas that are not usually seen by outsiders. In our schools these 'high status' areas are the entrance hall, the

headteacher's study, the main school hall, whilst 'low status' areas can include kitchens, storerooms and the caretaker's room.

## The School

| Room | Heat | Light | Space | floor | decoration | Walls | furniture | |
|------|------|-------|-------|-------|------------|-------|-----------|---|
| Head | †††　 | †††　 | †††　 | †††　 | †††　 | †††　 | †††　 | 21 |
| Staffroom | ††† | † †† | † | † | | † | † | 4 |
| Classroom | ††† | ††† † | † | † | †† | † | † | 12 |
| Kitchen | † | ††† | † ††† | † | | | | 8 |
| Boilerhouse | ††† | † | † | † | | † | | 7 |
| Cloakroom | † | † | | † | | † | | 4 |
| Corridoor | † | † | | | | †† | | 4 |
| Caretaker | † | † | | | | † | | 1 |
| Staff | ††† | ††† | †† | ††† | †† . | ††† | †† | 18 |

††† Lots or excellent
†† a lot or good
† a little or fairly good
0 = poor or nothing

**order of importance**
Head
Staffroom
Classroom
Kitchen
Boilerhouse
Cloakroom
Corridoor
Staff
Caretaker

## WHICH WERE THE IMPORTANT ROOMS?

The matrix is also a very useful tool for attempting to discover which areas of a site were important. Historic sites have areas which are of 'high' and 'low' status. They indicate a power or wealth differential. Whilst some areas might not be particularly 'rich', they take up a lot of space and relate to the specific function of the site, for example ramparts in a fort or castle, and in terms of that particular type of site they have a very 'high' status.

With domestic buildings that are still furnished in their original state it is easy

to detect the 'above stairs' areas where the owner lived and entertained, at least after the late eighteenth century, particularly in 'stately homes', and the 'below stairs' areas which were the workplaces, and living accomodation of the servants. Obvious clues to the more opulent areas will be wall coverings, large fire places, good quality furniture and objects. In structures that are partially or completely ruined, or have lost their furniture, decisions as to the status of an area are not so easily made. Clues, attributes, have to be gathered to form the basis for decision making. To make a decision about the importance

of a room or building, the simplest way of recording this information, is to present the data in an attributes or preference matrix. The first step is for the children to decide on the attributes they wish to use. At a castle, it may be the various levels of a keep they wish to categorise in terms of importance. Attributes might include the height of the rooms; the number of fireplaces, the richness of decoration around windows, doors, and fireplaces, the number and size of windows, etc. This would give a good guide to the status of a particular level, and it might be presumed that the higher a score for a particular level, the

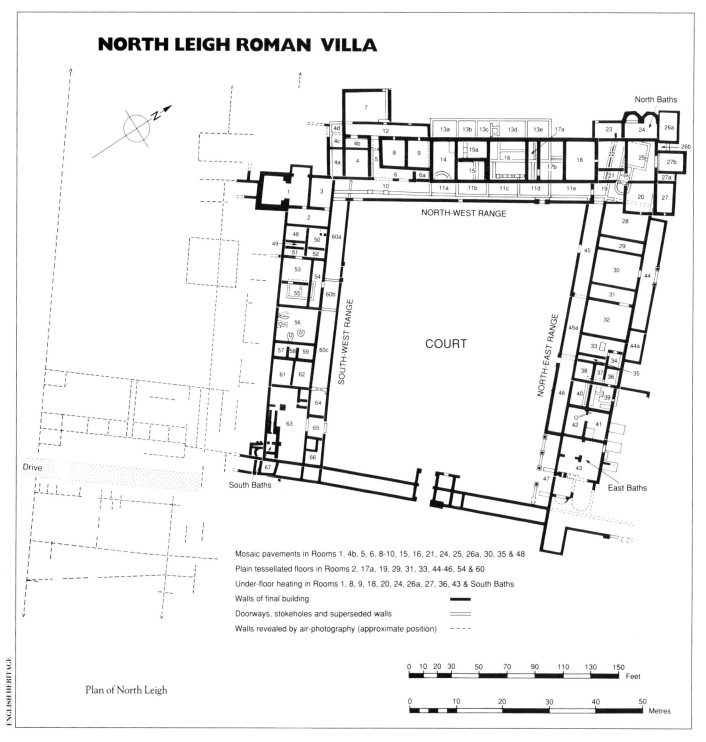

**NORTH LEIGH ROMAN VILLA**

Mosaic pavements in Rooms 1, 4b, 5, 6, 8-10, 15, 16, 21, 24, 25, 26a, 30, 35 & 48

Plain tessellated floors in Rooms 2, 17a, 19, 29, 31, 33, 44-46, 54 & 60

Under-floor heating in Rooms 1, 8, 9, 18, 20, 24, 26a, 27, 36, 43 & South Baths

Walls of final building

Doorways, stokeholes and superseded walls

Walls revealed by air-photography (approximate position)

Plan of North Leigh

ENGLISH HERITAGE

more important the resident was, or the function that took place there.

At a Roman villa that is is now reduced to footings or low walls, the size of rooms, the number of entrances, the quality of the flooring will give clues about the use of the room. At North Leigh Villa in Oxfordshire, a site that is mainly made up of low walls, children using the guide book, excavation plans, and the site produced a matrix to discover the previous use of the large number of rooms. The attributes in their matrix were the size of the room, its flooring, and whether it had under floor heating. They gave a high rating in a preference matrix to figured or patterned mosaic floors, which they thought would indicate great wealth, a medium rating to tesselated floors of large pieces or plain colouring, presuming that these might be in corridors or less important rooms, and a low rating to those areas covered in the 'opus signinum' type of concrete, flagstones, or earth floors, which they connected with kitchens, store rooms, or the servants quarters.

At Kenilworth Castle that had developed into a palace at a later date, children used a matrix recording the features they associated with the military use of a building, portcullis grooves, arrow loops, narrow windows, and the attributes they felt a palatial building would have: complex traceried windows, large, decorated doorways etc. They used the matrix to locate those

### Rochester Castle Keep

| Level | Heat | light | Space | floor | decoration | Walls | furniture | |
|---|---|---|---|---|---|---|---|---|
| 2nd floor | 8 | 8 | 9 | 0 | 9 | 0 | 0 | 34 |
| 3rd floor | 8 | 7 | 7 | 0 | 7 | 0 | 0 | 29 |
| 1st floor | 5 | 5 | 5 | 0 | 5 | 0 | 0 | 20 |
| ground floor | 0 | 3 | 5 | 0 | 0 | 0 | 0 | 8 |

0 = nothing
10 = Excellent

order of importance
2nd floor
3rd floor
1st floor
ground floor

**Attributes matrix for Rochester Castle keep.**

**Rochester Castle, Kent.**

parts of the building they thought might have been part of the military use of the building and those parts that were built as a palace. They were able to detect both periods of building, and find areas which they thought might belong to either period. Guesses about the 'mixed' areas were related to subsequent modification, or the lord's ideas about his power. One child thought that the builder of the palace might wish to retain the idea of a fortress in his new buildings as this might indicate his power and strength.

Individual sites have different areas related to status, and likewise sites themselves vary in their importance to each other. If several similar sites of the same period are visited as a project, it is possible to use a matrix to compare them, and to ascertain their relative importance.

For example, a project on castles might look at the size of the site, the number of towers, the height of its keep, its defences and whether a drawbridge was present etc. Domestic Roman sites can be categorised by the number of mosaics, the number of rooms, bath suites, rooms with underfloor heating and so on. The fortified buildings on Hadrian's wall - the forts, milecastles and turrets - can be similarly categorised, and a hierarchy of importance built up.

There is no reason why abbeys or priories of the same order of monks should not be treated in the same way, or the buildings of the rich in the medieval period and later.

**North Leigh Roman Villa, Oxfordshire.**

## HOW WERE THE PARTS CONNECTED?

Every structure which was used for defence, worship, or living in, had interconnecting parts, as do modern structures. Their relationship is usually shown on a plan. From either archaeological or documentary evidence, or both, or even from current usage, we may be able to work out the function of individual parts of a structure. It is then only a matter of working out how they were related. Maths uses 'network' theory to show the relationships between places, but this is complicated by the fact that many parts of an historical structure, particularly religious buildings, castles and palaces, would have only been accessible to those of a particular rank or status.

## WHERE WOULD THEY HAVE SPENT EACH DAY?

It is possible to reconstruct the lives of those in the past and their individual territories and routes between territories. For example, the life of a Roman soldier stationed at Housesteads Fort on Hadrian's Wall would have been centred on the barrack block, the baths, the wall itself, and perhaps the settlement outside the fort, where his family might have lived or where he went for a drink. It is less likely that the headquarters building or the commandant's house would be part of his daily round, although he might have gone to the granary to pick up supplies when it was his turn. So, a network can be drawn for that soldier. Similarly the daily round of the fort commander can be plotted.

Within a monastery we know enough of the everyday life of a monk to draw networks for particular times of the day. Different religious orders would have had different rules. At Mount Grace Priory in Yorkshire the Carthusians rarely left their 'cells', but for the Cistercian monks at Hailes Abbey in Gloucestershire each day would have been marked by ritual routeways depending on the time of day and the services they would have been involved in.

In the Elizabethan house only certain areas were the 'territory' of the servants, and others usually the 'territory' of the lord. So again it is possible to construct the spheres of influence of certain individuals in the household. These can be developed as networks, showing where individuals might have spent their days, or as shaded areas showing the usual territories of people of certain rank.

Depending on the age of the children, a similar sort of exercise can be carried out for most sites. Younger children can trace routes taken by individuals on the ground, and draw their own 'route' pictures. This can be extended to networks, like those illustrated here, with older juniors.

## Which part of the site?

The connections between parts of a site forms a firm basis for introducing direction. The language of position can develop from simple descriptions, such as 'next to', 'above', 'below', 'right', 'left', 'near', 'far', to routes around the site can be given with left and right turns, or simple compass bearings. These are of great use if the children wish to develop their own guide books. More complex bearings, using a compass accompanied by distance, can introduce surveying techniques. Likewise, the children can grid the plan in a guidebook, and describe position through the use of coordinates.

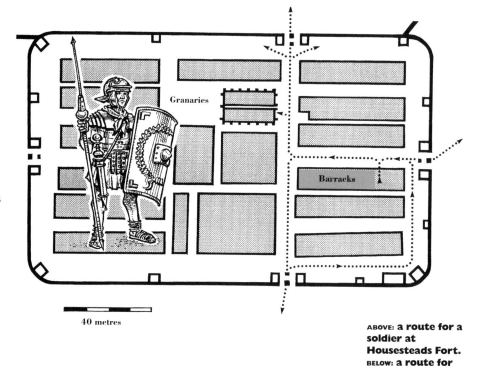

Granaries

Barracks

40 metres

ABOVE: **a route for a soldier at Housesteads Fort.**
BELOW: **a route for the Commander at Housesteads Fort.**

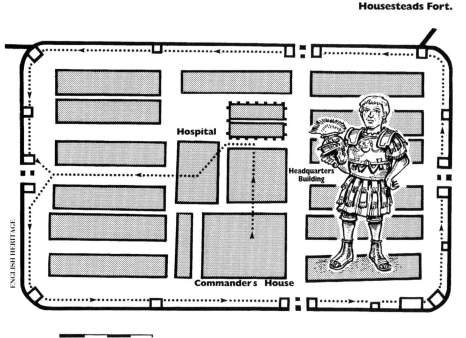

Hospital

Headquarters Building

Commander's House

ENGLISH HERITAGE

40 metres

## Hailes Abbey

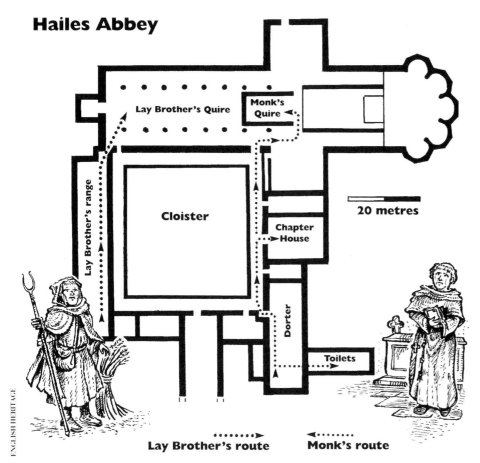

Lay Brother's Quire

Monk's Quire

Lay Brother's range

Cloister

20 metres

Chapter House

Dorter

Toilets

ENGLISH HERITAGE

•••••••► Lay Brother's route    ◄••••••• Monk's route

## HOW WAS THE PLACE CONNECTED TO OTHER PLACES?

An important part of visiting any site is to put it into its national, regional or local context. Even with very young children some idea of the distance away from school, even if it is expressed in hours, or 'from start of school until playtime' terms is necessary. With older children plotting the route on a map, calculating the time and distance to the site, looking at the type of relief on a map, or using a route flowchart, helps to put the site into its spatial context. It is important to remember that, in the past, time between sites was of equal importance to distance, and military or civil sites in upland areas needed to be closer together that those in flatter areas, simply because the terrain took longer to traverse.

ENGLISH HERITAGE

**Castle Acre Priory Norfolk.**

## HOW DID THEY GET FROM PLACE TO PLACE?

Most cultures have relied on a number of structures to form networks for defence, or the distribution of goods, and few sites after the Roman period stand alone. Before the Roman period it is less easy to be sure about the inter-dependence of places because of the lack of evidence, although long distance trade routes for trading stone tools, salt and pottery are known to have existed.

The Romans built up a series of defended forts and later a series of towns which themselves had connections with villages. Luckily the major sites were linked with roads which are largely known, and it is possible to see those links in use. The roads themselves were probably laid out with a groma, and the use of beacons, which explains their straight lengths. It has also been suggested that homing pigeons were used, which could have been released from either end of the proposed road, or at points in between and the angles of their flight paths recorded. Most substantial Roman sites give an indication of the road network that served them and joined them to the towns, and in the case of one we have documentary evidence for the goods carried into that site, although archaeology gives much information about routes and the goods carried, often outside the province of Britannia.

We know enough about the heirarchy of the elements of Hadrian's Wall to show how the turrets, milecastles, forts, and supply bases were connected.

From the surviving records of the itineraries that medieval monarchs undertook, and the routes of a variety of manufactured goods to the sites of sieges, for example, we are able to build up networks of the routes taken.

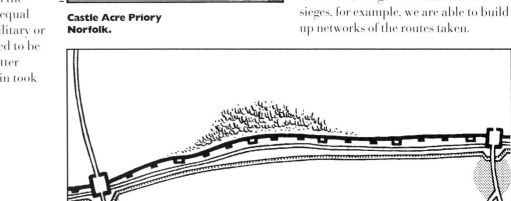

**Part of Hadrian's Wall showing the milecastles and turrets between two forts.**

ENGLISH HERITAGE

▬▬▬ The wall    ☐ Milecastle
▭▭▭ Road        ▪ Turret
▨▨▨ Ditch

# HOW BIG WAS IT?

The historic environment offers a multitude of opportunities for developing measuring skills in all forms, although weight is more confined to objects. Whichever stage the children are at, there will be useful problem solving work to do. There is a sequence of skill acquisition, which is related to children's abilities to use the increasingly complex tools of maths this involves:

■ establishing the boundaries and continuity of measurement;

■ comparison without the use of units;

■ using non-standard units;

■ the introduction of standard units;

■ extending the system.

Children will go through these stages at different rates, but all can be catered for on site.

## ESTABLISHING BOUNDARIES AND CONTINUITY OF MEASUREMENT

**Length:** 'Can you see where the wall begins and ends?' All visible historic sites have walls or earthworks and offer the opportunity for young children to trace them from Here to There, and to show the width of wide or narrow walls in the same way.

**Height:** 'Can you point to the bottom and then to the top?' This must be handled in much the same way with the children pointing from the base of a feature to its top, or in the case of dwarf walls, actually feeling their height.

**Area:** This is difficult for young children, but pointing out the extent of the area of a room that is now covered in grass, gravel, or flagstones, but bounded by walls can be useful in beginning to build up a concept of space covered. Using the hand to feel the area of different sized stones, or small

ENGLISH HERITAGE

windows can achieve the same result. 'Can you show me the floor of the room and where it stretches to, and what covers it?'.

**Volume:** 'What sort of thing could you do in this space?' Another difficult task, but just pointing out the 'space' within rooms: 'Is it big?', will be a useful start to concept development.

**Angle:** 'Can you use a part of your body to show the turning of the wall?' Historic sites contain a multitude of angles between walls, and provide children with the opportunity of getting into a corner and turning their heads, legs or arms through the angle. If doors or windows survive, then the angle of opening can be compared with body movements in much the same way.

## COMPARISON WITHOUT THE USE OF UNITS

**Length:** 'Which is the longest? Which is the shortest?' The language of length and width, longer than, shorter than, wider than, can be used on many parts of sites where nearby walls can be compared for their length or width characteristics.

**Height:** 'Which is the tallest/deepest?' Walls on most sites are differentially

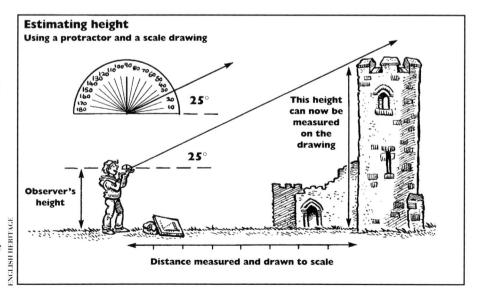

**Estimating height**
**Using a protractor and a scale drawing**

25°

This height can now be measured on the drawing

25°

Observer's height

Distance measured and drawn to scale

ENGLISH HERITAGE

ruined, and make a unique laboratory for comparing their heights, and guessing whether the remains stand to there full height, or are the result of falling down for some reason. The depth of wells, or moats can be measured by using string with a weight attached, and if a fresh length of string is used on each occasion, the results can be compared.

**Area:** 'Which room is the biggest?' This concept remains difficult using the large areas of rooms, although very different sized rooms, or areas of courtyard can be searched for the biggest and smallest.

**Volume:** 'Which room could you put the largest thing in?' Similarly rooms can be examined to see which 'Has more space than' another. This is particularly interesting with rooms of different shapes, tall thin towers, compared to low wide basements.

**Angle:** 'Which walls turn through more than a right corner?' The angles of walls, particularly those standing only a few courses, can be compared by deciding whether they are larger, smaller, or equivalent to a right angle. This will be made much easier by the use of a large cardboard right angle, prepared before the visit.

## USING NON-STANDARD UNITS

**Length:** 'How many ...... is it long?' Historic sites really come into their own with this sort of activity, as many of them were laid out using just this sort of informal, relative, measure. Pacing, foot length, arm length, span, cubit, hand span can all be used to measure the length, or width of walls, ditches, moats (across the bridge!) and courtyards. The perimeter of circles, their diameters and radii are equally suited to this sort of activity.

**Height:** 'How many .... is it high?' Perhaps the best way of using non-standard units is the use of the body against a wall, or earthwork, and the estimation of how many body lengths high, or deep, that feature is can be gauged by using fingers, or a stick and thumb. However, thumbs, or finger lengths at a distance are equally as valid. Stairways up castle mottes, or on to the top of ramparts, are useful tools for measuring relative height by counting the number of 'risers' ( the

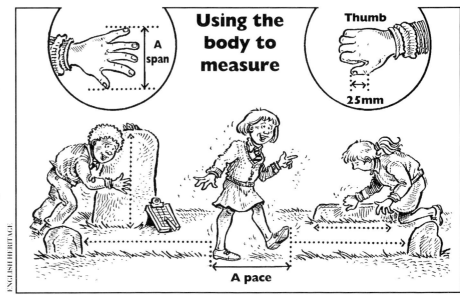

**Using the body to measure**

A span

Thumb

25mm

A pace

ENGLISH HERITAGE

vertical section between each 'treader'). Counting the number of courses of stone, where the stone courses are about the same thickness.

**Angle:** 'How many cut-outs does the wall turn through?' Any cut-out angle can be used to compare the amount of turning between two walls. So a wall angle is so many of the cut-out angles.

**Area:** 'How many .... cover the surface?' The human body is an excellent informal measure of area in spaces, especially where those spaces are known

to be used for sleeping (How many lying down?) or to move around comfortably. Many medieval sites are covered in tesselating tiles or flagstones, and again these make useful informal indicators of area. The individual tesserae of Roman mosaic pavements are really too small to be used in this way.

**Volume:** 'Does it have more space than our classroom?' Explorations comparing volumes of spaces and perhaps asking the questions: 'What would fit in here?'

MIKE CORBISHLEY

**Grimes Graves, Norfolk. James provides an informal measure for one of the filled-in pits of the neolithic flint mines.**

ENGLISH HERITAGE

**Kenilworth Castle, Warwickshire. Using a span to measure the width of a wall.**

# USING STANDARD UNITS

**Length:** 'How many units is it long?' Measuring walls, tracks, distances between one building and another, with the centimetre and metre, using trundle wheels, metre sticks and measuring tapes.

**Height:** 'How many units is it high?' Measuring low walls in the same way, using tapes to measure the heights of walls by dropping it down the wall surface, measuring the height of the risers and adding the results (using a calculator for extensive runs) to find the total height of the feature.

**Angle:** 'What angle does the wall turn through?' Using protractors and clinometers to measure angles taken by pathways, or between walls.

**Area:** 'How many square..... is the surface?' Measuring using the square metre, exploring the area of polygonal rooms on the site. Is there any general rule that can be invented?

**Volume:** 'How many cubic metres is the space?' Working out the volumes of small, accessible rooms with a cubic metre.

# EXTENDING THE SYSTEM

**Length:** 'How many ..... do you think it is?', 'Which would be the best unit/tool to use?', 'Can we draw it smaller or larger using a different scale?' Measuring more accurately, using estimation, comparing the accuracy of the measuring tool. Drawing plans to scale. Scaling up and down.

**Height:** 'Can we find lots of ways of finding the height?' Using clinometers and shadow sticks. Using hachures, the 'tadpole-like' signs that have the top of a slope at the body end, and the bottom of the slope at the tail end, with the length of the body denoting the length of the slope, to represent height.

Linear measuring systems used in the past can be challenging for older children, particularly when used on a site contemporanious with those systems. Little is known of the measurement system used by the pre-Roman peoples of Britain, but work on stone circles has suggested a measurement named the Megalithic Yard: Megalithic yard = 0.829m.

# MEASURING SYSTEMS

## Roman

| Denomination | Factor | Length | |
|---|---|---|---|
| | | Metric | Imperial |
| Pes (foot) | unit | 29.6 cm | 11.65 in |
| Digitus (finger) | 1/16 | 1.85 cm | 0.73 in |
| Uncia (inch) or Pollex (thumb) | 1/12 | 2.47 cm | 0.97 in |
| Palmus (palm) | 1/4 | 7.4 cm | 2.91 in |
| Cubitum (elbow or cubit) | 1 1/2 | 44.4 cm | 17.48 in |
| Passus (pace) | 5 | 1.48 m | 4.86 ft |
| Decempeda (10 feet) or Pertica (rod) | 10 | 2.96 m | 9.71 ft |
| Actus (furrow) length | 120 | 35.52 m | 116.54 ft |
| Stadium (stade) — 125 paces or 1/8 mille | 625 | 185.00 m | 606.9 ft |
| Mille passus (1000 paces) — the Roman mile | 5000 | 1480.00 m | 4856.00 ft |
| Leaga (league) — 1500 paces | 7500 | 2220.00 m | 7283.00 ft |

We know much about the Roman measurement systems because of their use across the Empire, and records detailing them. The unit in Roman linear measurement was the 'pes' which was normally of 29.57 cm. It was often referred to as the 'pes Monetalis' because the standard was kept in the Temple of Juno Moneta in Rome. Another 'pes' found on Romano-British sites is the 'pes Drusianus' of 33.3 cm. This was derived from the tribes of Germany who used this measurement as a 'foot'.

## Saxon

| Denomination | Factor | Length | |
|---|---|---|---|
| | | Metric | Imperial |
| Northern foot | unit | 33.5 cm | 13.2 in |
| Palms | 1/4 | 8.38 cm | 3.3 in |
| Thumbs | 1/12 | 2.79 cm | 1.1 in |
| Barleycorns | 1/36 | 0.93 cm | 0.37 in |
| Land Rod | 15 | 5.03 cm | 198.00 in |
| Furrow length | 600 | 201.17 m | 660.00 ft |

The Saxon measuring system was based on the 'Northern foot'.

## Norman

| Denomination | Unit | Length |
|---|---|---|
| | | Metric |
| Barleycorn | 1/108 | 0.847 cm |
| Inch | 1/36 | 2.54 cm |
| Foot | 1/3 | 30.48 cm |
| Yard/ulna | unit | 91.44 cm |
| Rod | 5.5 | 5.03 m |
| Furlong | 220 | 201.17 m |

The Normans allowed this system to remain intact, but in 1305, Edward I defined the basis of the Imperial system of measurement used until the advent of the metric system in Britain.

## Elizabethan

In the reign of Elizabeth I the mile is added:

| Mile | = | 1760 yards | = | 1.61 km |
|---|---|---|---|---|

There is a lot of problem-solving including estimation and arithmetic that can be generated by these comparative measurements. 'Which set of measurements would the builders have used?' 'What are the dimensions in that set?', 'Did they use them accurately?'.

**Area:** Comparing small areas with large. Using the hectare. Calculating irregular areas, and areas within areas (the different parts of a mosaic compared with its whole). 'How can we find the total area by adding the smaller sections?'. At this stage of the concept development of area it is possible to count the number of tesserae in a square metre of a mosaic, and to work out the approximate number of pieces in the whole, or the number of stones in a wall.

**Angle:** Castles are particularly good places for further work with angles. Medieval England was a place of war and the long bow was the main weapon, used by both attackers and defenders. Any archer knew that if he aimed his arrow horizontally at a distant target, he succeeded only in driving his arrow into the ground a short distance away. If the arrow was to travel a distance of more than 20 metres, then it would have to be aimed upwards. The flight path was approximately parabolic. A particular angle of elevation would give the arrow its maximum range. If the angle of elevation was too great, then the arrow might fall short and return to the ground, at an angle close to the vertical, and possible inflict injury on those members of his own side in front of him.

Cross winds were also a problem, and had to be allowed for. Usually the archer had some simple device for detecting wind strength, such as a piece of cloth on a stick set in the ground near him. The speed of the wind would have allowed the archer to judge how far he would have to aim to allow for the wind direction. So the archer in the open had to make decisions based on the mathematical data he learned to evaluate instinctively. For today's pupil the experience of actually taking those decisions can be a fruitful source of mathematical problem solving. Arrows can be fired to model the archer's task, and angles of elevation recorded. If this cannot be attempted, then at least it should be possible to calculate distance and direct angle of elevation to a target

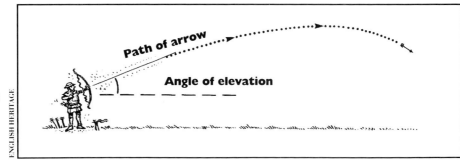

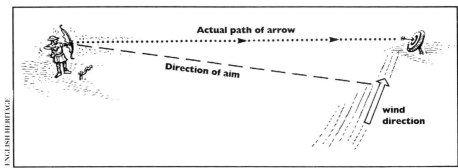

such as an arrow loop in a wall. A cross bow bolt would have taken such a flat trajectory.

The archer inside a tower of a castle, or on a wall, had a more complex task. He was constrained by the stone around him, and arrow loops, or the crenelations on top of the wall, had to be designed to give him enough room to manoeuvre. We tend to find deeply splayed spaces behind the loop to let the archer, or his 'spotter', allow for the wind and the position of his target. On top of the walls there would have been no problem with seeking an angle of elevation, but within the loop itself there would have to be space to enable an archer to get the height he needed.

On site children can use intact arrow loops and crenelations to guage the archer's field of fire, and the angles associated with it. Measuring the inner width of the apperture, its depth and the width of the outer opening will give enough measurements to draw a plan of the loop and to then determine the angles of its walls. A clinometer can be used to guage the vertical angle of fire from eye height. In recent experiments the arrow tended to wobble slightly before getting the velocity to fly straight, so it was necessary to stand back from the loop. How does this effect the field of fire horizontally and vertically? The cross bow had no such problem, because of the increased velocity provided by the tauter bow string. What would be the effect of this on field of

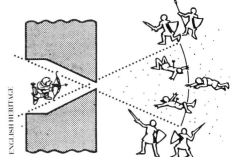

**Measuring the vertical angle of fire from an arrow loop.**

fire? Which type of arrow loop gave the maximum field of fire for which weapon?

Another way of determining the field of fire from an arrow loop is to position children on the ground outside the castle and have them move until they are at the limits of sight from a particular loop. If these positions are marked by string it should be possible to work out the angle of fire from the loop.

A similar exercise can be carried out when comparing fields of fire from a number of loops on a straight curtain wall to detect overlap, and to repeat the exercise from towers positioned in front of the wall. A straight section of curtain wall gives a field of fire infront of the wall that has a number of 'blind spots'. Do these areas disappear with the building of a round tower standing out from the wall, and by how much? What about polygonal towers? Do they increase fields of fire?

Children can also use angles to plan parts of buildings by using triangulation. Triangulation involves

**Cruciform slit for crossbows.**

**Splayed slit for longbows.**

having three measurements in order to produce a triangle. One of these measurements is the baseline, either a fixed feature, or a measured line indicated by two pegs. In the illustration AB is the baseline, and from either point

a particular feature, C, is measured using a tape. On the plan AB can be drawn to scale, and using compasses with the arcs the scaled measurement of AC and BC, the rest of the triangle can be constructed, and so indicating the

position of C. Of course the clinometer can be used horizontally to guage angles, and in the appropriate shaped structure, the use of right angled triangles can also be used. The present height of walls can be determined with a clinometer and tape measure using the right angled triangle created by the wall and the surface of the ground. It may be possible on some sites to determine the actual height of the wall. This can be done where a 'scar' of the fallen wall is left on an intact structure using the same process.

Where a ditch is exposed in front of a Roman wall the original height of that wall can be discovered. The shape of the ditch was usually determined by the height of the intact wall. The ditch should not provide protection for attackers, so its interior had to be visible from the top of the fort wall. This usually meant that its profile on the wall side was at an angle that allowed vision into it.

## VOLUME

'Which ...... has the largest volume? Why?' Working out volume of rooms, towers etc. The area and/or volume of a room can sometimes indicate its status. Larger rooms were usually more important, but other indicators are usually necessary.

■ On prehistoric sites it is possible to work out the weight of the individual stones through a fairly long, but simple method.

■ Whilst at the site measure the width, height, and depth of the part of the stone that can be seen above ground.

■ When back at school use a piece of local stone, or preferably a piece of stone picked up near the prehistoric site, but NOT from it, to work out its volume, by displacement (as illustration) and weigh it.

■ Work out the average volume of the prehistoric standing stone (Height × Depth × Width).

■ Then divide the volume of the standing stone by the smaller sample:

So the weight of the standing stone will be X multiplied by the weight of the sample stone.

# HOW OLD IS IT?

The task of putting the site into its context in time is one of the most difficult, not necessarily just because evidence is hard to find. Finding a date for a structure or an artefact is really about seeking out the relationship between the thing, or place, and the individual, between the past and 'me' and 'now'. It is about trying to get things in proportion, and that sense of proportion is related to the notion of how much things have changed. Time is a measure of how things have changed. When we describe something as 'timeless', we are in fact saying that it has not changed at all.

We can measure time in two ways. We can give a measure in terms of actual years, days, hours, minutes, seconds, or we can say that one event is earlier or later than another event. The ways of looking at time past are called chronologies, and so there are two types of chronology: absolute and relative.

## CHRONOLOGIES

Absolute chronologies are made by using actual dates: 'the date of this building is 1234', or 'the date of this coin is 1988'. In order to be 'absolute' we need either to have a date inscribed on some item, or for that item to be found in the same 'context' as something we can date accurately. A relative chronology is built from relating one event, object or structure to another, where we know that one is earlier or later than another - 'this Viking ship is earlier than this galleon', or 'this galleon is earlier than this speedboat'. On some sites documents give dates, and so we can date them absolutely, but other sites, particularly those before the Roman period, are often dated relatively, although with techniques using radio-active decay of the elements, such as wood or bone, a date can be given to some sites.

Fortunately, the idea of relative dating and absolute dating is equivalent to children's development in relation to using dates, where the 'relative' is similar to the 'informal dating' of young children, and the 'absolute' similar to

**Stonehenge, Wiltshire.**

the formal dating of 'years elapsed.' The task with children is to fit the use of dating on site to their growing understanding of the concept of time. This, again, comprises:

■ establishing some sort of boundary for time and expressing its continuity.
■ comparing times without the use of standard units
■ using non-standard units
■ using standard units
■ having an appreciation of a wide number of measures of time and dating.

## WHERE WOULD THEY HAVE BEEN?

Young children are very egocentric, and the boundaries and continuity of the measuring of time past are related very much to a short period before the present. With children of the ages five and six years, using any sort of dating system is difficult, and perhaps inappropriate at an historic site. Work in the classroom and the local

environment needs to be related to the events of their own lives and the events they have taken part in. This will be the seed of getting a simple notion of time passing, and the development of the concept of change. Although we can use statements such as 'a long time ago' with young children, it is perhaps best to try to relate the events of their lives with the everyday events that took place on a site, and explain that although people ate, played or fought on particular parts of a site that has now fallen down, they did basically the same things that today's children do, and had the same sorts of families. This can be emphasised by showing the children the parts of the site that would correspond to a daily cycle of activity, such as places for sleeping, places for eating, and places for work. It is important that they actually do some of those things in the appropriate place to make the connection between past and present.

Comparing time without the use of standard units is difficult when the children have such little experience of time elapsed. Since you cannot 'see' time, it is hard to say that one stretch of time is longer than another at an early age. More than anything we try to extend the boundaries of the six-eight year old's concept of time, and by doing so try to show the concept of the continuity of time.

| WHAT HAPPENED BEFORE...? | |
|---|---|
| **Before human beings** | **After human beings** |
| **The Earth, the sea, air, rocks, grass, flowers, rivers, fossils, reptiles, dinosaurs, water, clouds, the wind.** | **All the things they see on site that were made by human beings.** |
| **A curious facet of children's acquisition of a sense of time is their ability to realise that fossils, and dinosaurs and other 'prehistoric' creatures existed in the far reaches of the past before human beings appeared. Rocks** | **and fossils are a major interest to young children, and can be capitalised on in developing very basic chronologies. Again using a simple matrix, this can be recorded, often using the children's suggestions.** |

## WHICH ORDER DO THEY GO IN?

The use of a sequence of familiar actions, the day to day happennings of a child's life, recorded using a flowchart can be extended to the construction of familiar buildings, by visiting a building site if possible, and the home or school, if they have been extended at some time.

Clear the land
|
Dig trenches
|
to put the walls in
|
Build the walls and put the doors and windows in
|
Put the roof on
|
Make the paths and gardens
|
Paint the rooms
|
Move in

It is then a short step to investigating the building sequence of a particular type of ancient monument on site, for example a Norman motte.

Clear the land
|
Raise a fence to defend the builders
|
Heap the motte up and form a ditch around its base
|
Build a wooden Tower on the motte. Make a larger fence to keep the animals in
|
Move in

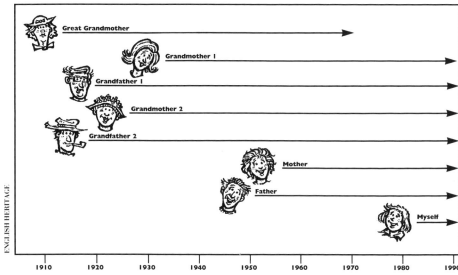

ENGLISH HERITAGE

Great Grandmother

Grandmother 1

Grandfather 1

Grandmother 2

Grandfather 2

Mother

Father

Myself

1910  1920  1930  1940  1950  1960  1970  1980  1990

Or in the case of a prehistoric stone circle:

Find a site that is flat
|
Mark out a circle
|
Bring the stones
|
Dig holes for the stones
|
Put up the stones

## WHO LIVED WHEN?

Such tasks will give a sense of sequence and enable more complex relative chronologies to be attempted. These are best developed by working backwards from the present, and using familiar objects. So a particular site was in use before motor cars, or aeroplanes were invented, before grandmothers and grandfathers were born. Such simple comparisons help the introduction of chronologies to children of six and seven years.

There needs to be a visual way of 'seeing' time, just as a number line is used in the classroom. The use of a 'sequence line' is essential, where particular items or events can be recorded in relationship to each other in terms of 'before' and 'after', leading to 'earlier than' and 'later than'. Again the events of their own lives will take prominence, and these can be extended further to events before they were born, but with little emphasis on trying to quantify these in any numerical way way. This can then be developed into a simple 'time line' with a scaled distance to indicate 'a long time ago', 'a short time ago' or 'the present' and can give a basic concept of time past, especially if pictures of castles or Roman soldiers can be used, and recorded in relative distances along the line. This is a very simple relative chronology. The presence of the evidence of Roman forts, castles and other monuments in the present landscape will underlie the idea of the continuity of time passing.

## WHEN DID IT HAPPEN?

Young children can be very perceptive in the use of 'certain ways of living' types of timelines with a 'naming scale'. Such 'typologies' of the standards of past life can act as non-standard units of measurement, and can be compared to their own development, their family relationships, the development of means of transport, anything they are familiar with. Usually by the ages of seven years or eight years, children have developed a 3 stage chronological framework in which to put such developments - 'Long Ago', 'In History', and 'Nowadays', and this is useful to relate other developments and sites visited. Such

| WHAT HAPPENED BEFORE...? | |
|---|---|
| **How they lived before I was born**<br>  **They didn't have taps or baths**<br>  **They used big fires**<br>  **They used arrows** | **How I live today**<br>  **I have a bathroom**<br>  **I have radiators**<br>  **We use guns** |
| **A further refinement of this simple form of relative, or informal chronology when on an historic site is to record the evidence for the way of life before** | **they were born and compare it with what they experience themselves. The process of change is central to this sort of exercise.** |

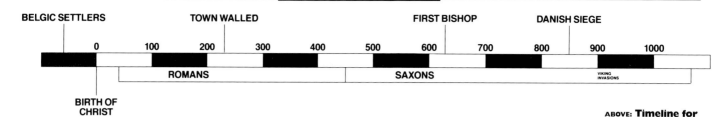

ABOVE: **Timeline for Rochester Castle.**
BELOW: **A site flowchart.**

language as 'Long, long ago', 'very, very, old', 'very old', 'old', 'fairly old', 'modern', 'recent', 'brand new', can be related to 'millions of years', 'thousands of years', 'hundreds of years', 'a few hundred years', 'last century', 'this century', 'within their own lifetimes' or 'a few years', and 'this year'. Such relative sequences can be used to place the sites visited in a more complex context, with children looking for clues that shed light on the 'ways of living' that happened at the location, but it is still a relative chronology.

## WHICH DATE?

As children's growing understanding of our denary number base extends so does their conception of time past, especially with their own informal time scales to relate to at the same time. Coming to terms with the full implications of historical dates and our system of reckoning time usually begins at about eleven years of age.

However, the length of a period as short as a century is still very difficult to grasp, and family timelines using an absolute chronology can still be used with older junior children in the study of more recent history, with local buildings being included on it. Another good introduction to absolute time scales is the graveyard survey, where the stones provide dates of birth and death, names, occupations and homes, but this should only be attempted with children fluent in the use of our number system.

As children's formal time sense develops to centuries, AD and BC, it is still useful to have an informal scale as part of the time line, which is becoming more sophisticated perhaps, with more

detailed typologies included (types of building, types of object - tools, vehicles, toys), but it is always worth remembering that particular types of object were used at the same time as more sophisticated ones, depending on culture, location, and status of social groups. A further refinement of this process is to examine a number of familiar structures for their architectural and building characteristics, such as window and door shapes, and to get the children to put them into their supposed relative order, then to look for foundation stones to find the absolute chronologies of the structures, or in a church to look at the dates of tombs, and to see if that compares with the dates of additions. Many schools have the dates of additions, and the name of the benefactors, on plaques, giving the necessary absolute dating.

On an historic site the guide book can be used to find the often relative dates of particular buildings, or additions to those buildings and these can be recorded by date and styles of building and materials so that a flowchart can be built up. However there is often a premise that because a building is in ruins it must be older than one that is intact. Plans in guidebooks and models showing the sequential growth of the site, either borrowed or made by children, are useful aids to understanding.

The flowchart can be used on a site to record the absolute chronology of a structure, and the relative chronology of the buildings. It is a compact vehicle for interpretation, recording vertically rather than horizontally.

| | |
|---|---|
| Site begun | 1086 |
| Timber Keep on a motte | 1088 |
| Bailey added | 1100 |
| Keep replaced in stone | 1120 |
| Stone curtain wall added to bailey | 1150 |
| Hall built in bailey | 1200 |
| Site falls into disrepair | 1250 |
| Site disused | 1320 |
| Site becomes an ancient monument | 1926 |

Parallel flowcharts can indicate the chronologies of other sites visited, and with national events, can be built into a network. With upper juniors it is possible to use the development of some types of buildings to show change over a wider context than just the individual site. Castles and defensive structures are frequently visited and their histories are firmly bound up with their shapes being altered by technological change. Hence square towers give way to round towers as a result of the development of stone throwing machines and the ability of a round form to throw off missiles. Attention is focussed on the gateways, resulting in elaborate forms at the expense of the keep. Towers appear on the walls, and a number of concentric walls are used to combat siege towers. These changes can be shown in a list or using a flowchart.

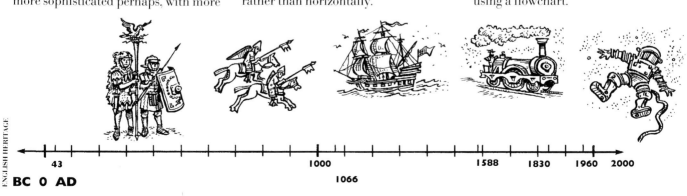

Both the flowchart and timeline indicate the linear nature of time passing. Often 'time circles' are used with the '12 o'clock' position indicating the present and the very distant past. These give the idea that time is cyclical, which in our culture it evidently is not, and if any representation besides the line is used perhaps the spiral, with a number of years, or the seasons in each 'loop', is the most accurate.

Children will also become familiar with the idea of dates getting smaller as the approach the Christian era , and a fluent use of the BC/AD convention, supported by a flow chart. The use of the AD/BC convention is also a useful introduction to negative numbers.

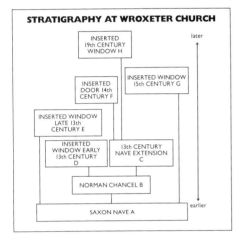

**STRATIGRAPHY AT WROXETER CHURCH**

**Recording the Stratigraphy of the building in matrix form.**

## WHAT IS ABOVE AND BELOW?

The use of relative dating is not something to be confined to the early years, when children are only able to use informal methods of dating. Relative dating techniques are an invaluable and sophisticated way of dating aspects of buildings for which there is no way of determining their absolute date and are demanding of older children.

The process of stratification is concerned with the incremental growth of buildings, or layers of earth. The school, if it was not built all at one time is a good example of horizontal stratification. By comparing the materials and styles, and possibly foundation stones, a flow chart can be built up showing which part came first and the date, if it is known. In the case of older structures, whilst they have spread over the years in a horizontal fashion,

they are likely to have been altered many times by rebuilding, especially, as in the case of town buildings or parish churches they have only a limited amount of space into which they can expand. So, in a more complex manner, it is possible to detect changes in materials, stone size and shape, mortar lines, that indicate a new stage of building activity. The best way of recording this after identifying the 'sets' of building activity, and justifying them, is to make a sketch of the wall, number the individual building blocks, and to use a flowchart to put them in time order. There may be datable objects within a block - a coat of arms, a date, a set of the initials of the owner - which will enable the dating of a particular block, and therefore all above it being younger than that date, and all below it being older.

**Parish church at Wroxeter, Shropshire.**

ROGER WHITE

ENGLISH HERITAGE

LEFT: **Remains of a Roman Villa at Little Oakley, Essex. The section shows a cut through a pit dug to dispose of rubbish.**

MIKE CORBISHLEY

BELOW: **Recent excavations at Wroxeter Roman City, Shropshire, exposing the remains of buried buildings.**

MIKE CORBISHLEY

Stratigraphy can also be seen below ground and many guidebooks and excavation reports, or museum exhibits, have illustrations of the stratigraphy in the section cut by archaeologists. Different activities make up different soils.Some of the levels are dark through having rubbish, straw, ashes, in them, so showing signs of habitation. Some levels are empty of signs of inhabitation, and their soil is made up of the remains of rotten leaves and other types of vegetation, and likely to be a cleaner and more uniform layer. As with the building blocks, such layers have boundaries, and when drawn can be flowcharted, to show the sequence of their deposition. Sometimes datable finds in the layers - coins, pottery, inscriptions, for example, can give an absolute date to a level.

Another use of relative dating techniques is to develop more sophisticated typologies relating to the development of buildings, castles, abbeys, and to artefacts, for example pottery forms, found in them.

It is well to remember that not all the sites in this country were built in the Christian era, and used the conventional chronology that we are familiar with. We do not know the way Celtic peoples recorded time, but we are familiar with their festivals. Roman sites produce dates related to particular Emperors, and the years of their reigns, as do later sites.

Many sites which do not produce datable objects, particularly prehistoric sites, can be dated by the use of radio carbon dating techniques. These use wood and other carbon based materials whose carbon 14 (C14) elements decay at a fixed rate. By measuring the change in the C14, and working out how much of it is left, it is possible to produce probability estimates of dates ranges within which the likelihood of the site's existence can fall. The use of the 'normal curve' to express these findings will be a familiar one to children used to plotting shoe sizes or heights as the same sort of curve is the result of both exercises. For example, a study of shoe sizes in any class room will produce a 'normal curve', because that is the curve that results from random processes acting on human beings. An extension of the findings will show that if any individual is approached then the chances are that individual will have a shoe size corresponding to the highest part of their curve. As parts of the curve to the left and right are explored, the chances of an individual in that class having a shoe size in that range gets less. So with radio carbon dating. The most probable date for a site will fall in the middle of the range, but it could still be likely that the site's date lies either to the left or right under the curve. This is expressed in 'standard deviations', which record the chances of the date deviating from the 'norm', which lies at the highest part of the curve. So a date of 3000 BC +/- 100, means that the greatest probability lies in the 3000 BC range, but it could be 100 years either side of that but with a decreasing probability.

# WHY WAS IT SHAPED LIKE IT IS?

Shape is important feature on historic sites, and they display a wide variety of geometric figures.

## WHAT SHAPE IS IT?

Many of our prehistoric religious sites are circular, possibly based on the Sun or Moon. Woodhenge, Stonehenge, Mitchell's Fold are all based on a circular shape. Sometimes circles are also used where large numbers of people collect to see an event, such as the Roman Amphitheatre at Silchester. The circular shape also allows maximum vision out of a structure as well as in. The towers of medieval castles, such as Launceston, or the shell keep of Restormel, the shape of many of Henry VIII's coastal forts and the Martello Towers of Napoleonic times are circular for this reason as well as offering no awkward angles that are easily damaged. The circle is also the most efficient shape for bounding an area: maximum space is attained with minimum perimeter. This has obvious benefits for defensive sites.

Squares and rectangles are the most common shape in every age's building. The shape has strength at its corners to hold up stone work, and was used in the shape of most Roman forts. Sometimes there is a combination of round towers on square bases, giving strength, vision, and smooth profile to deflect missiles. The towers at Goodrich Castle are an example of this.

Triangular buildings are rare. One exception is the Triangular Lodge at Rushton in Northants. However, the triangle is frequently found in the timber roof frames of many buildings, because of its weight distributing properties. Triangular spurs were increasingly added to the forts of the sixteenth and seventeenth centuries as a means of deflecting cannon shots. The defences of Berwick-on-Tweed are a good example of this.

Polygons of some sorts are chosen for their strength in supporting the weight of large buildings such as towers. The keep of Conisborough Castle is hexagonal, as is the keep of Old

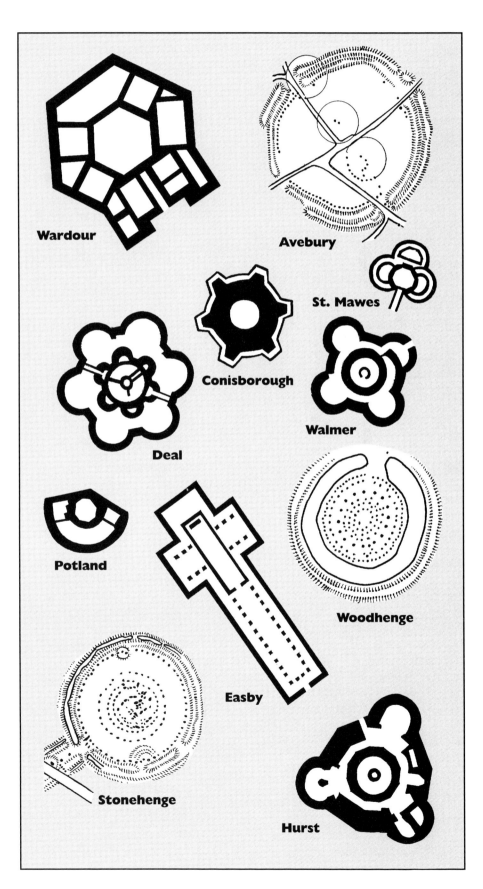

Wardour

Avebury

St. Mawes

Conisborough

Deal

Walmer

Potland

Easby

Woodhenge

Stonehenge

Hurst

Wardour Castle. The Roman lighthouse at Dover is octagonal. The chapter house at Thornton Abbey and the Yarn Market at Dunster are octagonal because of the need for as many straight surfaces as possible. Many other shapes are found in the decoration of houses of the Classical period, and the geometrical designs of Roman mosaics, and most sites composed of masonry rather than just earthworks will have a variety of shapes to identify and explain.

'How many different shapes can we find?' 'Why was that particular shape used?' 'Can we draw it accurately?' The shapes used in the ceilings of some classical buildings can generate some inventive art and design work, as alternative decorations are made.

## SYMMETRY

Human beings seem to have an inbuilt love of, and need for symmetry, which is reflected in the use of symmetrical forms for decoration in many buildings, both in their plans and elevations. Each of the shapes listed above has the property of symmetry, but it is their use together that gives buildings their symmetrical properties. There are many reasons for this, the shape of the structure having some relationship to its function:

■ the need for taking stresses equally

■ enabling equal defence on all sides.

■ enabling equal vision on all sides.

■ giving equal space on each side.

■ decoration

■ a combination of all of these.

Symmetry comes in two forms. Line or reflective symmetry, and rotational symmetry. Line symmetry is where one half of a building is exactly mirrored by the other half. Buildings can have one or more lines of reflective symmetry. A shape has rotational symmetry if, after rotation, every image point is mapped onto the object shape. The 'order' of rotational symmetry for a shape is the number of possible rotations, that give rotational symmetry. Although rarely found in buildings, rotational symmetry is often a feature of planned gardens in classical houses.

Tilbury Fort, Essex. The fort was begun in 1670. Its angle bastions were revetted with massive brick walls.

Arches of all sorts of buildings are symmetrically designed because they have to take equal weight from above. Where they are assymetrical it usually means that unequal weights are being taken, and it is presumed that this was a lesson learned in the early stages of the building attempts of humans. Windows are similarly symmetrical, and often their smallness is due to the difficulty of designing a building with large apertures which could take the weight of stonework. In Norman buildings we find that the thick walls could not be pierced by large windows, and it was not until the development of buttresses, particularly flying butresses, also placed symmetrical opposite each other, that we find that windows could become larger, yet still symmetrical.

From the earliest times people learned that the most efficient use of space, combined with the need for defence, was to build to a symmetrical plan, where the terrain allowed. The Romans arranged the barrack blocks symmetrically around a fort so the ramparts or wall could be manned equally to meet threats from any side, and to protect the buildings inside.

Similarly the other great use of symmetry for defence was in the concentric castles of the medieval period, seen particularly in Wales during the Edwardian period. Old Wardour Castle is hexagonal symmetrical, and so is Conisborough. The use of symmetry was also important in the coastal forts of Henry VIII, such as Deal, or Walmer, and for similar reasons.

Aligned to the need for defence, symmetry was also invaluable in ensuring equal vision on all sides, hence the often symmetrical shape of toll houses so that travellers could be seen coming on all sides.

Symmetrical forms have been important in the planning of buildings, both to give efficiency of space use, and

usually based on the rectangle. This is the case in all periods of the past.

The use of symmetry is found in the mosaics of Roman sites, the windows of the great medieval abbeys, and the formal gardens of Elizabethan houses, in each case to make them pleasant on the eye. This is particularly so with the great houses of the seventeenth and eighteenth centuries, where balanced buidings were found visually satisfying, and the shapes copied from the Classical period of Greece and Rome.

Triangular Lodge, Northamptonshire.

Perhaps the most extraordinary use of symmetry is in the Triangular Lodge at Rushton in Northamptonshire, where everything is symmetrical and based on the triangle. The use of symmetry in this case has religious connations related to the concept of the Trinity.

Symmetry is found in the plans of buildings and groups of buildings, their elevations, and in the detail of windows, doors, etc. 'Symmetry Trails' can be the focus of an investigation in any building. 'Can we find symmetrical features in the building?' 'Is the plan symmetrical?' 'Which sort of symmetry is it?' 'Can we find the lines of symmetry?' 'Why was symmetry so important in this building?'.

# TRANSFORMATIONS

Many sites have decoration that forms a pattern which we can analyse. The decoration helps us to decide which buildings or parts of buildings were special in some way, either as places where important people lived, or places that had an important religious function.

At many Roman villas important rooms, usually used for eating or entertaining, were given mosaic pavements. These mosaics were made up of small pieces of stone called tesserae that fitted together. They tessellated, completely covering the available space. The tesserae themselves made up a variety of patterns which also tessellated. The LOGO sequences below are for two patterns found on Roman mosaic pavements. At many sites the mosaics have an interlacing border. This particular pattern was taken from the "Venus and Cupids as Gladiators" mosaic at Bignor Villa in Sussex.

**Detail of a mosaic floor at Bignor Roman Villa showing Venus and cupids as gladiators.**

TO VENUS
HT
RT 135
FD 70
REPEAT 100 [FD 1 LT 1]
LT 75
FD 35
LT 100
REPEAT 50 [FD 1 RT 2]
FD 65
LT 100
FD 71
RT 90
FD 35
RT 90
FD 70
REPEAT 100 [FD 1 RT 1]
RT 75
FD 35
RT 100
REPEAT 50 [FD 1 LT 2]
PU
LT 180
REPEAT 50 [FD 1 RT 2]
FD 35
LT 135
PD
VENUS
END.

This program will repeat itself endlessly because of the repeat instruction 'VENUS' before 'END'. It is an interesting challenge to change the program to produce a right angled turn for each corner of the pavement.

**The Sea God mosaic floor from Roman Verulamium.**

The famous 'Sea God' mosaic from St. Albans, now in Verulamium Museum, produces a meandering pattern using squares and rectangles. It involves three separate program steps.

TO X//RETURN
REPEAT 2 [FD 150 RT 90]
REPEAT 2 [FD 65 LT 90 FD 40 LT 90
REPEAT 2 [FD 20 RT 90] FD 60
REPEAT 2 [RT 90 FD 20] LT 90 FD 40
LT 90 FD 65 RT 90] PU FD 20 RT 90
FD 20 PD
REPEAT 4 [FD 110 LT 90] PU FD 130
LT 90 FD 25 LT 90 BK 20
PD//RETURN
END//RETURN

This forms the corner piece of the border,

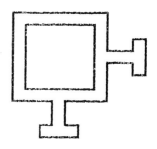

**It is possible to draw the shape of the border of the Sea God mosaic using Logo in three steps.**

```
TO F//RETURN
REPEAT 2 [LT 90 FD 45 LT 90 FD 200
LT 90 FD 45 LT 90 FD 20 LT 90
FD 25 RT 90 FD 30 RT 90 FD 110 RT
90 FD 30 RT 90 FD 25 LT 90 FD 20]
PU LT 180 FD 70 RT 90 FD 25 PD
REPEAT 2 [LT 90 FD 60 LT 90 FD 110]
PU LT 90 FD 150 LT 90 FD 130 RT 90
PD//RETURN
END//RETURN
```

This forms the side of the pattern.

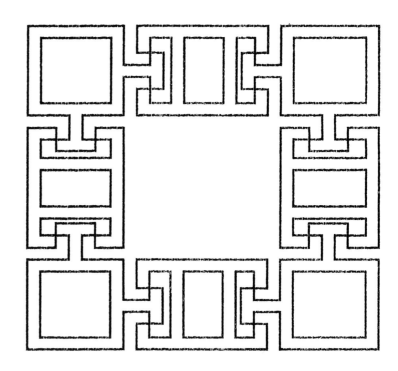

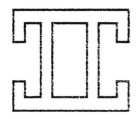

```
TO BORDER//RETURN
PU LT 90 FD 400 RT 90 FD 200 PD
REPEAT 4 [X F]//RETURN
END//RETURN
```

This forms the complete border.

In the original an interlacing pattern underlines the mosaic on one side. How could this be inserted in the program (using VENUS)? Could the 'Sea God' mosaic be given an interlacing pattern on all sides? Both sequences have elements of symmetry, particularly the 'Sea God' mosaic, and within the symmetrical design are 'transformations', and these underlie the notion of symmetry in their shapes and repetition. The transformations - translation, reflection, rotation and enlargement are part of the mathematical description of symmetry.

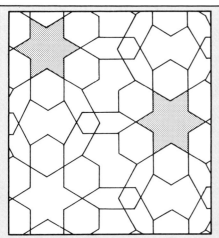

**TRANSLATION: is a transformation in which a shape slides without turning. Every point moves the same distance and the same direction.**

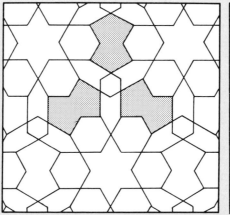

**ROTATION: is a transformation in which every part turns through the same angle about the same centre. When rotated the shape of a figure is not changed, but its position and orientation do change.**

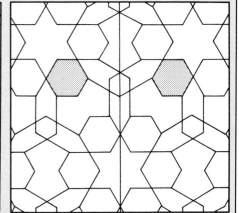

**REFLECTION: is a way of transforming a shape as a mirror does, a mirror image.**

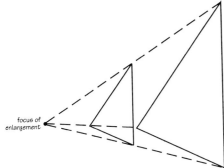

focus of enlargement

**Enlargement:** For translation, rotatation, and reflection the one constant is that the size of the figures remains the same. With enlargements it is the size that alters but the shape remains the same.

Transformations can be identified in mosiacs, in the patterns on ceilings in classical houses, and particularly in the plans and facades of Classical and Palladian country and town houses. 'What sorts of transformation can we identify?' 'Can we design our own pavements, or facades using transformations?'

Tessellation is often found in less grand buildings in the construction of a pavement with a pattern of regular polygons - triangle, quadrilaterals, etc. The shapes are usually translations but not always. 'Which shapes will tessellate to form a pavement?' 'Which transformation is used to tessellate particular shapes?' 'Can I design my own pavement using several tesselating shapes?' 'What makes shapes tessellate?'

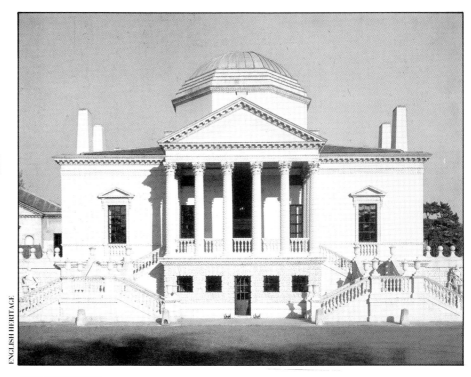

ENGLISH HERITAGE

**Chiswick House, London. A masterpiece of Palladian architecture, built c1727-9.**

ENGLISH HERITAGE

**Detail from the ceiling of the Blue Velvet Room at Chiswick House.**

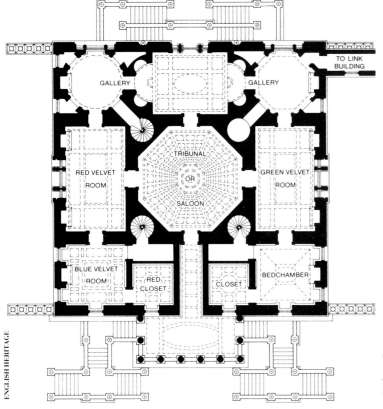

ENGLISH HERITAGE

**Chiswick House, London. The first-floor plan of the house.**

31

## HOW WERE THESE STRUCTURES PLANNED?

Much of the maths we use today is the result of thousands of years of development, and it is difficult for us to appreciate the simple maths of the neolithic and bronze ages when people built geometrically designed stone monuments and carved intricate patterns on stone. Although many of the resulting shapes look sophisticated, they are all the product of very simple techniques that can be used in the playground, or on a site with great effect.

Spirals were sometimes used to decorate carved stones in neolithic

ENGLISH HERITAGE

tombs. If we try to make a LOGO program to draw spirals we soon find how complex the maths is, yet they are simple to draw. Take a post and fix it into the ground and tie one of a string to

it. Fix a large piece of chalk to the other end and walk around the post, keeping the string taut and let it wind up as you walk. The chalk will draw a spiral on the ground surface, and it will gradually get smaller until all the string is wound tight. The creator of this shape in neolithic times perhaps didn't realise that it increases its diameter by the same amount each revolution, or that the distance betwen the lines is equal to the circumfernce of the central post, but the shape needed was formed nevertheless.

John Wood, who has explored these ideas, has examined the shapes of many stone circles and how they were designed, and it is his work that is being

## THE GOLDEN RECTANGLE

There is often more to the shape of buildings than straightforward symmetry. Structures of the Palladian and Neo-Classical period often exhibit a shape whose parts have a particular ratio to each other, and is the result of the relationship between certain sets of numbers. This mathematical sequence is known by the nickname of its discoverer, Leonardo ('Fibonacci') daPisa. The Fibonacci series is produced by starting with 1 and adding the last two numbers to arrive at the next: 1,1,2,3,5,8,13,21,34, etc. The effect of these numbers is seen powerfully in Nature with the spirals of a daisy head having a ratio of 21:34, the pine cone's spiral 5:8, and the bumps on a pineapple, 8:13.

Fibonacci numbers also appear to exert a strange influence on architecture. The ratio between any two Fibonacci numbers after 3 is about 1:1.6. This is known as the Golden Section, or the Golden Ratio. The ratio - expressed more precisely as 1:1.618 - occurs in pentagons, circles and decagons - but notably in the Golden Rectangle, a figure whose two sides bear the magic relationship to each other. The Golden Rectangle is said to be one of the most visually satisfying of all geometric forms and has been used in buildings that were imitating the classical architecture

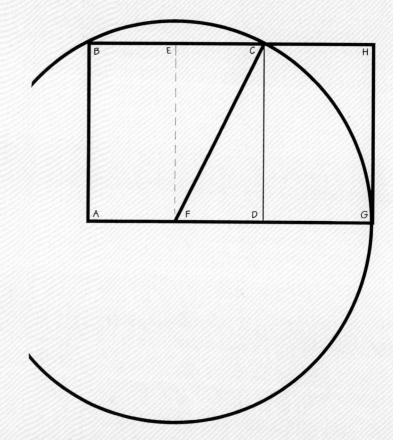

of Greece and Rome where it was frequently found.
To construct a Golden Rectangle, you need to begin with a square ABCD, which is then divided into two equal parts by a line EF. Point F now serves as the centre of a circle whose radius is the diagonal FC. An arc of the circle is drawn (CG) and the base line AD is extended to intersect it. This becomes the base of the rectangle. The new side HG is now drawn at right angles to the new base, with

the line BH brought out to meet it. The resultant Golden Rectangle can be used as a template on a larger sheet of paper. If it is drawn around, and the Golden Rectangle shape cut out, an easy to use 'Golden Rectangle detector' is made that can be taken to any site to see if the shape influenced the configuration of the building. Some buildings use the Golden Rectangle vertically as well as horizontally, so both directions need to be investigated.

used here. Most of the stone rings known are in the form of a true circle. Stonehenge and the North and South Circles at Avebury are 'true' circles. The simplest way of laying out this type of circle on the ground is to use a length of string or rope twice as long as the radius of the circle to be formed. One end of the loop is put over a post fixed into the ground and to the other end is fixed the marking material. The circle is drawn by walking around the post keeping the string tight all of the time. A loop is much better than tieing a single piece of string to a post as it is less likely to snag.

If a 'true' circle is being visualised, it may be possible to find the centre of the shape, by crossing two pieces of string from opposite sides, and to test the circularity with a tape. Using string, which stretches greatly, is not as accurate as using thick rope. Neolithic builders may have used grass ropes or hides which would have stretched if they got wet. A further problem would have been undulating or uneven ground. How might they have overcome these difficulties?

Many stone rings use an ellipse, for example Winterbourne Abbas in Dorset, or Postbridge in Devon. This shape is also fairly simple to draw. An ellipse is a circle with two centres, amd is drawn in a similar way to the circle but with two foci, pegs or stakes in the ground, a loop of rope over each. The further apart the stakes are, the more flattened will be the ellipse. The shape was probably discovered by accidental snagging when trying to draw a circle. There are many more complex shapes, the easiest of which is probably Alexander Thom's 'Flattened circle: Type B', an example of which is at Merrivale in Devon. It is made by:

■ drawing a complete circle with a loop of rope turning about a post at O.

■ fixing another post into the ground at a point on the circumference of the cicle at A.

■ measuring a right angle from OA and laying out line MN. By folding the rope into thirds finding the points C and D and hammering in two more posts.

■ taking a loop of rope so that it goes over the post at C, and its length is just right to bring the other end of the loop to M.

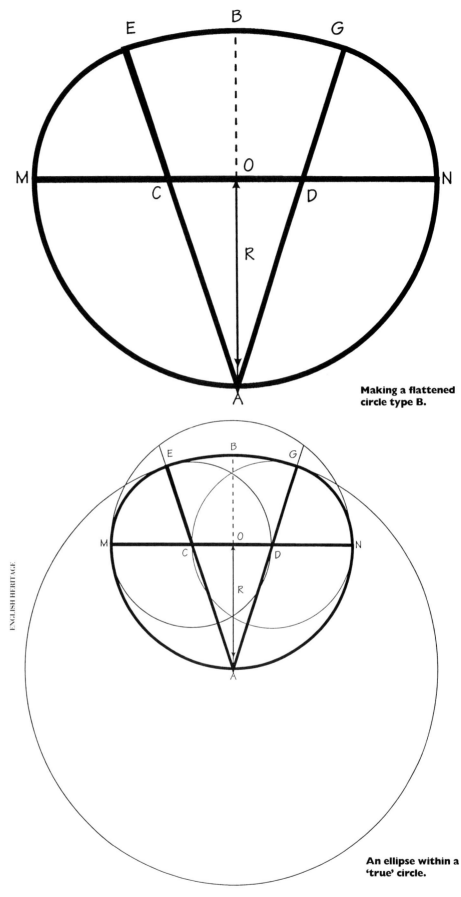

**Making a flattened circle type B.**

**An ellipse within a 'true' circle.**

■ putting a marker in the loop and swinging it around so that it first draws a circle centred at C then as it clears the post at C it will begin to draw a circle of radius AE.

■ continuing the movement as far as G, then letting the loop pivot about D to complete arc GN.

## ROMAN SURVEYING

Whilst we know little about the surveying methods of prehistoric times, we are fortunate to have a detailed knowledge of Roman surveying techniques. The main emphasis of Roman planning was the square or rectangle, and so the right angle became the crucial element. Having decided where to site the fort or town, the surveyors laid out two principal axes, the 'cardo maximus' and the 'cardo decumanus'. When these were in place it was possible to plan the rest of the town. In order to establish the right angle a surveying instrument, the groma, was used. This consisted of a cross frame with four arms set at 90 degrees to each other with a plumb line suspended from the end of each arm. The instrument was supported upon a staff which was positioned off centre to enable the surveyor to sight through the opposing plumb lines and survey straight lines and right angles. Another line hung from the centre of the frame so that it could be set up over the exact point on the ground. Poles, known as metae or signi were used to sight straight lines, whilst linear measures were made with graduated ten foot rods (decempedae), footlines or with strings and markers.

## WINDOWS AND ARCHES

During the medieval period the use of geometry was fundamental in the development of the arch and windows. The architect or builder started from the laying out of the window or arch before building it in stone. We have to use the finished product and try to work backwards.

Finding the centre of a Roman or semi-circular Norman arch is straightforward. We need to join the two ends of the arch with a straight line, and then take the mid-point of this line as the centre. The radius of the arch can then be drawn.

The Norman arch is very simple because the builders did not pierce thick walls with large windows or doors. With the development of architectural engineering skills, particularly the use of external butresses to take the weight of the roof, walls became thinner, and it was possible to create large and complex openings such as windows. Complex Gothic windows are not so easily reconstructed, and actually designing one indicates the level of skill

**Using the Roman groma.**

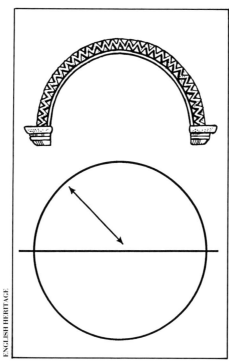

**A semicircular arch would have used a full circle as its base.**

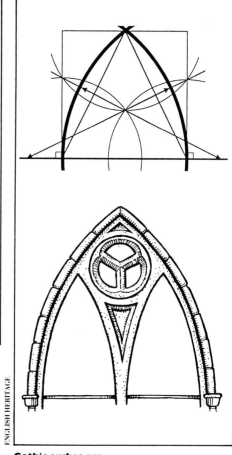

**Gothic arches are more complex.**

needed by the original architect.

To find the centres of the two arcs that make up the arch, we have to assume that the tangent to each arc is perpendicular to the line of the base of the window, and that therefore the centres of the arcs lie on this line. The centres will also lie on the perpendicular bisectors of the lines which join the ends of each arc. The two centres we are looking for lie where the perpendicular bisectors intersect the line of the base.

# WHAT ARE WE SURE OF?

| 0 | 1 | 2 | 3 | 4 |
|---|---|---|---|---|
| **No chance** | **Poor chance** | **Even chance** | **Good chance** | **Certain** |
| | **(unlikely)** | **(possibly)** | **(probably)** | |

Trying to understand historic sites is like grappling with a variety of unknowns. We are always faced with a limited repertoire of evidence in trying to explain the whys and wherefores of any site. This is particularly true the farther back in time that we travel. The builders of prehistoric sites didn't leave any 'time capsules' to record their intentions, and many of the sites that we meet with are 'prehistoric' in their lack of the recorded word to explain the events and thinking that went into the construction of particular buildings and other edifices. That is one of the reasons that the past is so interesting we are left to interpret the evidence ourselves. We examine the evidence and have to balance the probabilities. Throughout the history of archaeology interpretations of sites have changed dramatically, not always with new evidence being forthcoming, and there is no reason why children should not examine the available evidence and try to work out the probabilities themselves. Of course there is always the danger that adults will try to impose their own interpretation, or the current received wisdom, on the child. We all have to start somewhere in developing the skills of gauging probabilities, and childhood, or young adulthood, is as good as any place to start.

The excavator of Gatcombe Roman villa in Avon actually set out the results of the excavation in terms of 'certain', 'near certain' and 'hypothetical'. We take a lot for granted, and historic sites are a good place to start thinking about the chances of things being 'true'.

What can be said to be 'sure' at any site? That the Romans built this place for a military purpose? That they lived there for a particular length of time? That they did particular things there? How do we know?

Conversely we visit a medieval castle, where the guide book might tell us that we have the records of the building, who built it, who did the work, and why, where and when it happened. The probabilities are greater in this case, although we still have difficulty in allocating rooms to particular functions, and have to resort to physical evidence that is not so 'secure'. Using basic probability ideas we can record findings on a scale from 0 -4.

So can we assign the categories of probably, possibly, unlikely to particular interpretations of a site. Children can use their own reasoning powers with the available evidence at a number of levels: the buildings alone, using the guide book, using the excavation report. They can make statements to reflect their findings. Not only is such an activity an interesting approach to probability theory, it also underlines the tenuousness of some of the knowledge we have of historic sites.

## HOW DO WE SHOW IT?

As we have seen, using a form of representation to aid understanding can be a useful way of collecting and displaying data. Throughout this book a wide range of representational formats have been used, and historic sites are excellent places to develop this sort of problem solving aid. The table below is a guide to the appropriate format for a particular purpose.

ENGLISH HERITAGE

| What you want to do | Type of representation to use |
|---|---|
| **Describe** | **Numbers, ordinal or cardinal scales** |
| **Classify** | **Topic web, Venn diagram, tree diagram, attributes matric** |
| **Compare** | **Bar chart, rating, attributes matric preference matrix** |
| **Order** | **Time line, time chart, ordered list, network, flowchart** |
| **Position** | **Drawing, scale model, map, scale drawing, plan** |
| **Interrelate** | **Table, histogram, scatter graph, graph** |

# BIBLIOGRAPHY

There are few books directly related to the use of mathematics and the historic environment and all are written at adult level.

## General

Orton, C, **Mathematics in Archaeology,** Collins, 1980, ISBN 0-00-216226-1

## Weights and Measures

Connor, R D, **The Weights and Measures of England,** HMSO, 1987, ISBN 0-11-290435-1
Dilke, O A W, **Reading the Past: An Historical Outline,** Yale University Press, 1965

## Stone Circles

Thom, A, **Megalithic Lunar Observatories,** Oxford University Press, 1971
Wood, J E, **Sun Moon and Standing Stones,** Oxford University Press, 1978, ISBN 0-19-211443-3

Each of the following English Heritage Handbooks for Teachers contains examples of the use of Mathematics on historic sites.
Barnes, J, and Hamsworth, A, **Dover Castle,** English Heritage, 1991, ISBN 1-85074-320-7
Brown, S, and Clutterbuck, R, **Totnes Castle,** English Heritage, 1988, ISBN 1-85074-200-6
Copeland, T, **Kenilworth Castle,** English Heritage, 1990, ISBN 1-85074-292-8
Copeland, T, **Rochester Castle,** English Heritage, 1990, ISBN 1-85074-238-3
Cooper, R, **Carisbrooke Castle,** English Heritage, 1988, ISBN 1-85074-238-3
Fairclough, J, and Redsell, P, **Living History: reconstructing the past with children,** English Heritage, 1985, ISBN 1-85074-310X
Meades, D, **Pevensey Castle,** English Heritage, 1991, ISBN 1-85074-309-6
Planel, P, **Old Sarum,** English Heritage, 1991, ISBN 1-85074-308-8

Pownall, J, and Hutson, N, **Science and the Historic Environment,** English Heritage, 1991, ISBN 1-85074-331-2

Other resources incorporating approaches to historic sites that include mathematics:
Aston, O, Primary **History Problem Solving,** Shropshire Education Department, 1989, Investigative techniques, includes castles.
Copeland, T, 'Problem Solving in History', **CBA Education Bulletin, No 3,** June, 1987
Copeland, T, 'Schools for Thought', **CBA Education Bulletin,** No 5, June 1988. Using the school as a starting point for studying buildings.
Copeland, T, 'Castles Topic Pack', **Junior Education,** August 1990. Ideas for cross-curricular approaches
Corbishley, M, 'Not much to Look at', **Remnants,** 1, pp 1-2, English Heritage, 1986, Beginning to look for evidence
Corbishley, M, 'The Case of the Blocked Window', **Remnants,** 2, English Heritage, pp 1-4, 1986, Above ground archaeology
Corbishley, M, 'Miss ... Please Miss, Why did People Live Underground?', **Remnants,** 3, pp 1-3, English Heritage, 1987, Using archaeological remains with children
Corbishley, M, 'Coping With the Bird's Eye View', **Remnants,** 8, pp 8-9, English Heritage, 1989. Understanding plans
Corbishley, M, 'Play the Skeleton Game', **Remnants,** 8, pp 8-9, English Heritage, 1989, Using evidence from graves
Corbishley, M, 'Play the Dustbin Game', **Remnants,** 9, pp 12-13, English Heritage, 1989, Making sense of rubbish

## ACKNOWLEDGEMENTS

We are grateful to the following people for help in producing this book:
Students at Cheltenham and Gloucester College of Higher Education for devising the LOGO programs; Carol Anderson and the Young Archaeologists Club branch in Oxfordshire; The table in 'How do we show it?' was adapted from material in the Open University's course 'Mathematics across the Curriculum'; David Coles for his helpful suggestions on the drafts of this book.

ENGLISH HERITAGE

# English ⌗ Heritage

The Education Service aims to provide teachers with as much help as possible. Our free booklet **Information for Teachers,** contains a list of all our sites together with many practical ideas for getting the most out of school visits, and our booking form. Our catalogue **Resources** is also available on request. For copies of the above, further details of our teachers courses and any other information about our Education Service please contact:
**English Heritage
Education Service
Keysign House
429 Oxford Street
London W1R 2HD
071-973 3442/3
Fax 071-973 3430**